Cover: *Detail from* The Recall *by Don Troiani* (Historical Image Bank)

CMH Pub 75–1

The Regular Army Before the Civil War

1845–1860

by
Clayton R. Newell

Reprinted by St. John's Press LLC
Alexandria, Virginia 22301
ISBN 978-0-9662986-2-8
2015
Center of Military History
United States Army Washington, D.C.
2014

Introduction

Although over one hundred fifty years have passed since the start of the American Civil War, that titanic conflict continues to matter. The forces unleashed by that war were immensely destructive because of the significant issues involved: the existence of the Union, the end of slavery, and the very future of the nation. The war remains our most contentious, and our bloodiest, with over six hundred thousand killed in the course of the four-year struggle.

Most civil wars do not spring up overnight, and the American Civil War was no exception. The seeds of the conflict were sown in the earliest days of the republic's founding, primarily over the existence of slavery and the slave trade. Although no conflict can begin without the conscious decisions of those engaged in the debates at that moment, in the end, there was simply no way to paper over the division of the country into two camps: one that was dominated by slavery and the other that sought first to limit its spread and then to abolish it. Our nation was indeed "half slave and half free," and that could not stand.

Regardless of the factors tearing the nation asunder, the soldiers on each side of the struggle went to war for personal reasons: looking for adventure, being caught up in the passions and emotions of their peers, believing in the Union, favoring states' rights, or even justifying the simple schoolyard dynamic of being convinced that they were "worth" three of the soldiers on the other side. Nor can we overlook the factor that some went to war to prove their manhood. This has been, and continues to be, a key dynamic in understanding combat and the profession of arms. Soldiers join for many reasons but often stay in the fight because of their comrades and because they do not want to seem like cowards. Sometimes issues of national impact shrink to nothing in the intensely personal world of cannon shell and minié ball.

Whatever the reasons, the struggle was long and costly and only culminated with the conquest of the rebellious Confederacy, the preservation of the Union, and the end of slavery. These campaign pamphlets on the American Civil War, prepared in commemoration of our national sacrifices, seek to remember that war and honor those in the United States Army who died to preserve the Union and free the slaves as well as to tell the story of those American soldiers who fought for the Confederacy despite the inherently flawed nature of their cause. The Civil War was our greatest struggle and continues to deserve our deep study and contemplation.

RICHARD W. STEWART
Chief Historian

The Regular Army Before the Civil War, 1845–1860

The fifteen years that preceded the outbreak of the American Civil War were eventful ones for the U.S. Army. After invading and defeating Mexico, the Army dispersed across the vast Western frontier undertaking a myriad of duties. It subdued American Indian tribes, explored and governed new territories, and generally worked to maintain peace. At the same time, it supported national development through mapping and engineering projects, grew in size, and undertook important steps toward modernization. While these activities did not fully prepare the Army for the cataclysm that was to come, they did provide opportunities for officers to hone their leadership skills under trying conditions. Many of the men who would become generals during the Civil War first tasted battle as junior officers in Mexico and the American West.

The United States Army in 1845

The United States Army in the early nineteenth century was a modest institution. Concerns over economy and the threat that a large Army might pose to liberty led many Americans to regard the Army as a necessary evil at best. Fortunately, the relative security afforded by vast oceans and sparsely populated neighbors created a situation in which the United States could survive with a small military establishment. By 1845, the United States had about 7,300 men under arms to protect a nation of nearly 20 million people and 1.8 million square miles of territory. In contrast, Belgium's proximity to the great powers of Europe led that country of just over 4 million people and 12,000 square miles to maintain an army of about 30,000 men. Even Mexico, a nation of 7 million people with whom the United States would soon be at war, maintained a regular army of over 18,000 men with another 10,000 militia on active duty. If Americans enjoyed relatively inexpensive security in peacetime, a disorderly and costly rush to mobilize the nation's untapped financial and human resources would be the inevitable price for unpreparedness should the nation go to war. Such had been the case when America went to war with Great Britain in 1812, and the upcoming war with Mexico would be no different.

Subsumed under the Department of War headed by a civilian secretary, the U.S. Army was divided into bureaucratic and combat elements. The bureaucratic side of the Army consisted of 259 staff officers, 17 military storekeepers, and about 450 enlisted men organized into nine staff departments or bureaus. Heading these bureaus were the Adjutant General, the Inspector General, the Paymaster General, the Quartermaster General, the Commissary General of Subsistence, the Surgeon General, the Chief of Ordnance, the Chief of Engineers, and the Chief of Topographical Engineers. These bureau chiefs oversaw the Army's logistical and administrative affairs in fairly autonomous fashion. Although they reported to the secretary of war, the bureau chiefs were jealous of their prerogatives, enjoyed direct relationships with Congress, and tended therefore to be difficult to control. Also reporting to the secretary of war was a commanding general, whose exact role vis à vis the bureau chiefs was ambiguous and fraught with tension. As for the Army's combat forces, these consisted of fourteen regiments—eight infantry, four artillery, and two dragoon. Underfunded, undermanned, and dispersed among many small and often remote outposts, the soldiers of the line lived and worked hard for meager pay. Few of the aging senior officers had a professional military education, while 30 percent of the enlisted ranks were illiterate. In a land bustling with economic opportunity, many Americans looked down on soldiers as shiftless individuals who donned the blue uniform either because they could not or would not engage in the industrious pursuits of normal society. Nevertheless, the rank and file were tough men, inured to the discipline and exactitudes of Army life, and led by an increasingly professional officer corps because of the efforts of the U.S. Military Academy at West Point, New York.

The Mexican War

If many Americans did not hold regulars in high esteem, they recognized their utility when war appeared on the horizon in 1846 (*Map 1*). Tensions with Mexico started ten years earlier, in 1836, when Americans living in the Mexican state of Texas rebelled and formed an independent country, the Republic of Texas. Although the United States government did not intervene in the Texas Revolution, it recognized Texas' independence and the boundary it claimed with Mexico along the Rio Grande River, neither of which were accepted by the Mexican government. Reflecting national sentiment that it was America's "Manifest Destiny" to expand across the North American continent, the United States government eventually moved to annex the Republic of Texas in 1845. When the Texas legislature voted to join the United States in July of that year, President James K. Polk ordered Brig. Gen. Zachary Taylor to advance into Texas as far as

The American Soldier, 1836, *by H. Charles McBarron*
(U.S. Army Art Collection)

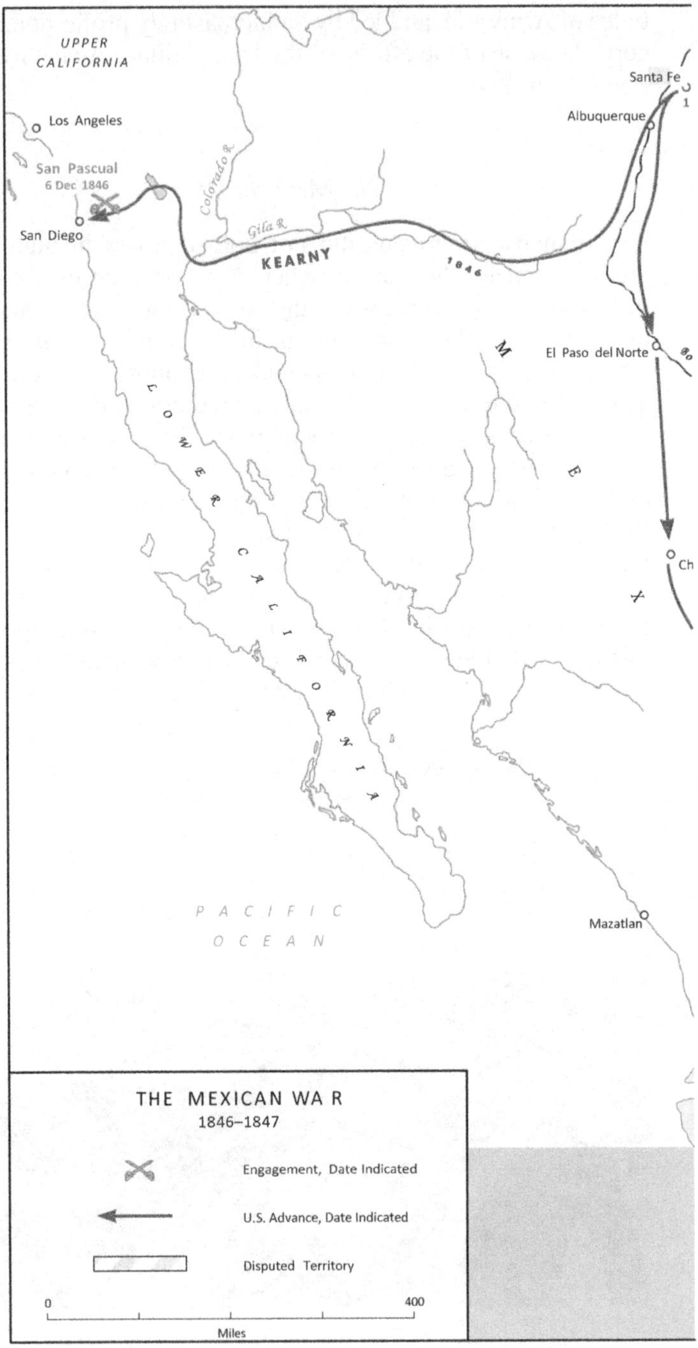
UPPER CALIFORNIA
Los Angeles
San Pascual
6 Dec 1846
San Diego
Colorado R.
Gila R.
KEARNY
1846
Santa Fe
Albuquerque
El Paso del Norte
LOWER CALIFORNIA
PACIFIC OCEAN
Mazatlan
THE MEXICAN WA R
1846–1847
Engagement, Date Indicated
U.S. Advance, Date Indicated
Disputed Territory
0
400
Miles

Map 1

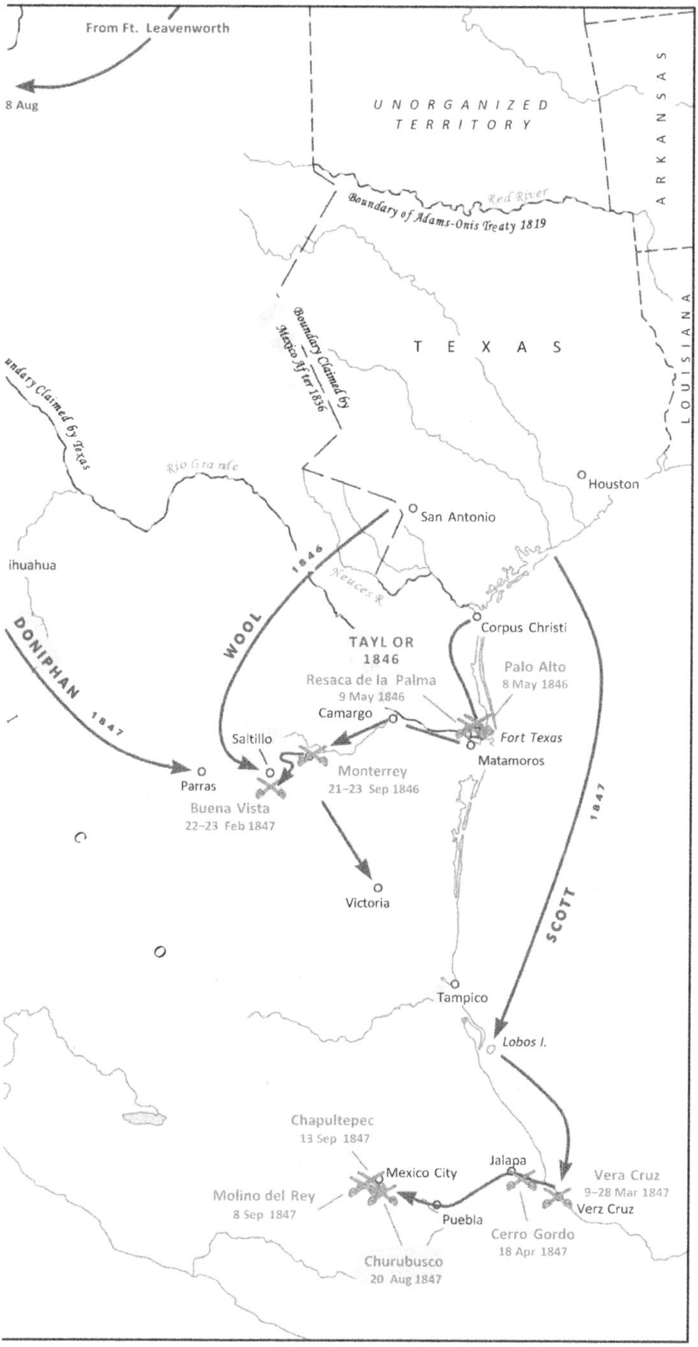

From Ft. Leavenworth
8 Aug
UNORGANIZED
TERRITORY
ARKANSAS
Red River
Boundary of Adams-Onis Treaty 1819
LOUISIANA
TEXAS
Boundary Claimed by
Mexico After 1836
undary Claimed by Texas
Rio Grande
Houston
San Antonio
ihuahua
1846
Neuces R
Corpus Christi
DONIPHAN
1847
WOOL
TAYLOR
1846
Palo Alto
8 May 1846
Resaca de la Palma
9 May 1846
Camargo
Fort Texas
Saltillo
Matamoros
Parras
Monterrey
21–23 Sep 1846
Buena Vista
22–23 Feb 1847
1847
Victoria
SCOTT
Tampico
Lobos I.
Chapultepec
13 Sep 1847
Jalapa
Mexico City
Vera Cruz
9–28 Mar 1847
Verz Cruz
Molino del Rey
8 Sep 1847
Puebla
Cerro Gordo
18 Apr 1847
Churubusco
20 Aug 1847

the Nueces River where Mexico claimed its border with Texas to be. By October 1845, Taylor had assembled 3,860 troops—about half of the entire Regular Army—at Corpus Christi, Texas. After negotiations broke down to get Mexico to acknowledge the Rio Grande as the proper boundary and to sell California to the United States, Secretary of War William L. Marcy ordered Taylor to cross the Nueces in early 1846 and to establish an outpost, named Fort Texas, on the Rio Grande River opposite the Mexican town of Matamoros. Taylor dutifully complied before returning with the majority of his men to Corpus Christi.

In late April and early May 1846, Mexican forces crossed the Rio Grande River, ambushed an American patrol, and laid siege to Fort Texas. Marching to break the siege, Taylor met the Mexicans at Palo Alto on 8 May 1846. His army consisted of elements of the 3d, 4th, 5th, and 8th Infantry, a battalion from the 2d Artillery serving as infantry, two squadrons of dragoons, and three batteries of cannons. During this first major engagement of the war, U.S. Army artillery played an especially central role. Battery A, 2d Artillery, commanded by Capt. James Duncan, and Battery C, 3d Artillery, commanded by Maj. Samuel Ringgold, took up positions in front. From their advanced positions, Duncan and Ringgold fired at the massed Mexican infantry, and because their guns outranged the Mexican artillery, the American artillery decided the battle. Artillery fire caused most of the Mexicans' estimated four hundred dead. In his report of the battle General Taylor wrote: "Our artillery, consisting of two 18-pounders and two light batteries (C, Third, A, Second) was the arm chiefly engaged, and to the excellent manner in which it was manoeuvred [sic] and served is our success mainly due." American losses totaled six dead and forty wounded. Among the dead was the dashing Major Ringgold, soon to be replaced by another brilliant artilleryman and future Confederate general, Capt. Braxton Bragg.

The day after the battle Taylor resumed his advance toward Fort Texas. At Resaca de la Palma, the Mexican Army blocked his advance once again, establishing a defensive line along a ravine. The defenses extended for about a mile on each side of the road with artillery in the center. Ulysses S. Grant, a lieutenant in the 4th Infantry who would go on to lead the Union Army to victory in the Civil War, commanded a company during the battle and recalled how the rounds fired by the Mexican artillery whistled menacingly overhead. When a company of U.S. dragoons overran but failed to hold the Mexican cannon, Taylor turned to Col. William Belknap, commander of the 8th Infantry, and shouted "take those guns, and by God keep them!" The regiment, accompanied by elements of the 5th Infantry, charged headlong into the ravine. After fierce fighting, the American infantrymen captured nine guns which they used to fire on the Mexican Army, bringing the battle to a successful end.

When word reached Washington that fighting had broken out along the Rio Grande, Congress moved to put the nation on a war footing. It appropriated $10 million to fund the conflict; created a company of sappers, miners, and pontoniers; and strengthened the Regular Army by increasing company strength from sixty-four to one hundred privates. It also created a new regular unit—the Regiment of Mounted Riflemen—to secure settlers moving west, although the administration would soon send this unit to Mexico as well. Last but not least, Congress authorized the president to call up fifty thousand volunteers to supplement the Regular Army. Unlike the regular units, the new volunteer regiments were raised, organized, staffed, and equipped by the individual states before being mustered into federal service.

After several months of accumulating supplies and incorporating some volunteer units into the Army, Taylor, now a brevet major general for his victories at Palo Alto and Resaca de la Palma, began marching toward Monterrey, Mexico, in August 1846. He reached the city the next month with about three thousand regulars and three thousand volunteers. Well fortified and defended by seven thousand Mexicans, the town of Monterrey posed a significant challenge. Taylor sent attack columns toward two sides of the city while artillery fired on the fortifications. The Americans encountered tough resistance as they pushed their way into the town. The Mexican commander offered to surrender on 24 September provided Taylor would permit him to withdraw unmolested and grant an eight-week truce. Taylor, believing generosity would further peace negotiations, agreed to the terms. However, when President Polk received the news, he terminated the armistice and directed Maj. Gen. Winfield Scott, commanding general of the Army, to conduct a campaign to capture Mexico City. Taylor did not welcome the news that most of his army, including virtually all of the regulars, would be reassigned to Scott in preparation for the upcoming offensive.

The departure of the regulars in January 1847 left General Taylor with fewer than seven thousand men, all volunteers except for a few companies of the 2d Dragoons and the cannon batteries from the 3d and 4th Artillery. Meanwhile, General Santa Anna, president of Mexico, assembled an army of some twenty thousand men. Taylor expected that Santa Anna would move against Scott when his army landed at Vera Cruz. However, when the Mexican general learned that most of Taylor's forces had been transferred to Scott, he decided to make a bold strategic move. He turned north hoping to win a quick victory over Taylor's smaller army. By 22 February 1847, the Mexican Army had fifteen thousand men arrayed against fewer than five thousand Americans near Buena Vista. When the battle opened Santa Anna hit Taylor's left hard causing it to collapse. As the frantic American retreat threatened to develop into a rout, Captain Bragg galloped onto the field at the head of Battery C, 3d Artillery. According to Taylor's report of the battle, the Mexicans were "but a few yards from the muzzles of his pieces. The first discharge of canister caused the enemy to hesitate, the second and third drove him back in disorder and saved the day." According to tradition, Taylor said, "A little more grape, Captain Bragg." However the 3d Artillery regimental history points out that in all probability the general really shouted "Captain, give them hell," as the guns raced into position. In any case, Santa Anna, having lost between fifteen hundred and two thousand men killed and wounded, decided to withdraw. Although a near run thing, the American victory ended the fighting in northern Mexico.

Just about the time Taylor began his march toward Monterrey in August 1846, the United States opened another front in the war. Col. Stephen W. Kearny, commander of the 1st Dragoons, left Fort Leavenworth, Kansas, with eighteen hundred men designated as the "Army of the West." Kearny's mission was to capture Santa Fe, Mexico, and then move on to occupy California. He had five companies of dragoons, two companies of artillery, and two companies of infantry from the Regular Army as well as nine companies of Missouri Volunteer cavalry. By 18 August Kearny had taken Santa Fe without a fight. Leaving the Missourians to hold the town, he moved on to California with the dragoons. By early December, as Kearny neared San Diego he met a small party of U.S. marines sent by Commodore Robert F. Stockton who had been directed to blockade Mexican ports and to assist Army operations in California.

An Army train crossing the plains
(Library of Congress)

On 6 December, Kearny attacked a force of Mexican lancers at San Pascual with two companies of the 1st Dragoons and the marines. Kearny's dragoons were mounted on mules tired from the long trek to California, and when the Mexican lancers charged they inflicted severe losses on the Americans. After their initial charge, however, the lancers withdrew and Kearny's forces occupied San Diego. Moving north, the Americans engaged 350 Mexicans just south of Los Angeles and quickly routed them, thus ending further resistance in California. In February 1847, the Missouri Volunteers who had been left in Santa Fe, along with a company of the 1st Dragoons, defeated a force of Mexicans south of El Paso. That action ended any further threats to the American conquest of New Mexico and California.

American successes in the north and west notwithstanding, Mexico refused to accept America's terms for peace. Realizing that the war would continue, in February 1847, Congress authorized the recruitment of ten new regiments for the Regular Army. The force included the 3d Dragoons, eight infantry regiments numbered the 9th through the 16th Infantry, and an unusual hybrid organization grandly designated the Regiment of Voltigeurs and Foot Riflemen. The unit was to consist of highly trained skirmishers—*voltiguers* in French—who would fight on foot but be transported by horses. Only half of the men in the regiment were to have horses. To facilitate short, rapid movements the man on the ground would ride behind the mounted soldier. In practice the concept failed to materialize and the voltigeurs operated largely on foot, as did the Regiment of Mounted Riflemen, which lost most of its horses in a storm while crossing the Gulf of Mexico in ships. What truly distinguished these two units from the rest of the regular infantry in Mexico was that they were armed with model 1841 rifles, weapons that enjoyed greater range and accuracy than the smoothbore muskets carried by most infantrymen during the war. Congress provided additional support in March when it increased the number of companies in each artillery regiment from ten to twelve.

While Congress initiated these measures in Washington, General Scott prepared his army for an amphibious landing at Vera Cruz on Mexico's east coast. From there he planned to drive inland to capture Mexico City. Scott set sail on 2 March with a force of more than thirteen thousand men that included both regulars and volunteers. On the evening of 9 March, he launched the Army's first major amphibious landing as more than ten thousand men went ashore in waves using sixty-five heavy surf boats. The troops landed unopposed on a beach approximately three miles from Vera

The American Soldier, 1847, *by H. Charles McBarron*
(U.S. Army Art Collection)

Cruz in about four hours. Pvt. George Ballantine, 1st Artillery, recalled the landings as "exciting and imposing." As he watched the first wave form up for the run to the beach, "military bands from different regiments stationed on the decks of steamers, transports, and men-of-war played . . . 'Yankee Doodle,' 'Hail Columbia,' and 'The Star Spangled Banner.'" The city quickly fell after a brief siege.

After capturing Vera Cruz, Scott advanced on Mexico City. His army met little resistance initially, but in April, when the army reached Cerro Gordo, it encountered a strongly entrenched Mexican force of about twelve thousand men holding positions in a rocky defile that controlled the road leading to Mexico City. General Santa Anna had emplaced artillery batteries on mountain spurs on the left of his position. On the right, Mexican guns positioned on El Telegrafo commanded the high ground that overlooked the highway. On the morning of 12 April, Mexican gunners opened fire on the advance elements of Scott's army while they were still out of range. Not wanting to assault the Mexican defenses directly, Scott sent his engineers to look for a route to outflank the position. Lt. Robert E. Lee, an engineer and future Confederate general, discovered a ravine that bypassed the main defenses. This route enabled Brig. Gen. David E. Twiggs to move his division of regulars into a position to attack the Mexican rear. Private Ballantine recalled his captain asking General Twiggs how far they should charge, and receiving the answer, "charge them to hell!" The general's response was prophetic as the fighting became a form of hell with both sides resorting to the bayonet in hand-to-hand combat. Pvt. Barna Upton in the 3d Infantry witnessed "a kind of fighting which I hope never to see again. It seemed like murder to see men running bayonets into each other's breasts." Twiggs' flanking movement succeeded in cutting the enemy's escape route, and the Mexicans surrendered in droves. Santa Anna lost between

Robert E. Lee in the Dress Uniform of a Lt. of Engineers *by William Edward West* (Washington-Custis-Lee Collection, Washington and Lee University, Lexington, Va.)

one thousand and twelve hundred casualties and three thousand prisoners. American losses were 64 killed and 353 wounded.

By the end of May 1847, Scott reached Puebla with just 5,820 effective enlisted men. Many soldiers had been hospitalized with wounds and sickness, and seven volunteer regiments whose enlistments had expired had departed for home with another 3,700 men. The general had no choice but to halt and await the arrival of fresh troops from the United States. After receiving reinforcements that brought his army to about ten thousand men, Scott resumed the advance on 7 August. With Mexican guerrillas blocking his supply line back to Vera Cruz, Scott needed to capture Mexico City before his supplies ran out. After three days of marching, the column reached a high plateau about fourteen miles from the capital.

Although the Americans were close to their objective, Santa Anna would not concede Mexico City without a fight. At the Churubusco River the Mexicans took advantage of the thick walls of the San Mateo Convent to conduct a determined defense that kept the Americans at bay for several hours. Second Lt. Henry J. Hunt, 2d Artillery, rushed forward, unlimbered his cannons under fire, and unleashed what Maj. Gen. William J. Worth termed "a fire of astonishing rapidity" at the fortifications. With the help of Hunt's guns, the 3d Infantry forced its way through the defenses, causing the Mexican garrison to surrender. The report written by the Mexican commander cited Capt. James M. Smith, a company commander in the regiment, as having been "the first to present himself upon the parapet," providing "an example of valor to many following him."

The Americans next met the Mexicans at Molino del Rey. Maj. George Wright, 8th Infantry, commanded the assault column. In his report of the battle he recalled: "At the distance of 200 yards the enemy opened on us with round and grape shot with considerable effect, the ground being perfectly level. I instantly ordered the double quick step; the line advanced rapidly and immediately came within musket range." After being hit by a musket ball and carried from the battlefield, Wright later reported that "the assaulting column continued in combat . . . until the enemy's positions were all carried and we remained in possession of the field; after which there being but three officers left and the rank and file very much reduced." The 5th Infantry conducted repeated charges against Casa Mata, a fiercely defended stone structure located five hundred yards from Molina del Rey. The 5th, along with the 6th and 8th Infantry met determined opposition and attacked repeatedly before finally driving the defenders from the position. The 5th Infantry paid a high price for its success. The regiment started the battle with 13 officers and 342 enlisted soldiers. It lost 3 officers and 27 men killed, 4 officers and 107 men wounded, and 5 men missing—some 40 percent of its strength.

The citadel of Chapultepec remained the final obstacle to capturing Mexico City. On 13 September, a storming party began the assault. The 4th Infantry regimental history succinctly describes the action: "Under a terrific storm of shot and shell the party reached the ditch and main walls of the fortress, scaling ladders were brought up and amid hand-to-hand fighting a lodgement [sic] was secured." Mexico City surrendered the following day. As the 3d Infantry led the victorious American Army into the Grand Plaza of the city, General Scott is reputed to have removed his hat and told his staff to "take off your hats to the Old Guard of the Army." The 3d Infantry, which today serves as the president's official escort and guards the Tomb of the Unknown Soldier in Arlington National Cemetery, Virginia, continues to be known as the Old Guard.

On 2 February 1848, the Treaty of Guadalupe Hidalgo officially ended the war (*Map* 2). The last American soldier left Mexico in August, and the national boundary between the two countries became the Rio Grande. At the same time, Mexico ceded the territories of California and New Mexico, thereby adding approximately 1.2 million square miles of territory to the United States. During the war the Army had demonstrated its toughness and discipline, and Scott credited his victory in Mexico to his battle-hardened soldiers. The nation mobilized about one hundred sixteen thousand men for the war, although the number in the field at any one time was significantly smaller. Over forty-two thousand men served in the Regular Army, whose five-year term of enlistment meant that most of them were in uniform for the entire war. The service of the mobilized militia and U.S. Volunteers varied more widely according to the terms of their enlistments. Some served for three months, others for twelve months, while still others stayed for the duration of the war. Of the approximately seventy-four thousand men who entered volunteer service, only about thirty thousand served in Mexico.

Although the vast majority of men who fought in the Mexican War were raw recruits, they benefited by the existence of the most professional officer corps since the creation of the U.S. Army. By 1847, the U.S. Military Academy, founded in 1802, had produced

WESTERN FRONTIER
1817–1848
Boundary Fixed by Treaty with Spain, 22 February 1819
Territory in Dispute with Mexico, by Texas and later United States
U. S. Army Forts
0
100
200
300
Miles
CANADA
MICHIGAN
Fort Snelling
1819
Fort Atkinson
1840
Chicago
Fort Atkinson
1819–1827
Mississippi R.
ILLINOIS
INDIANA
Fort Leavenworth
1827
Fort Belle Fontaine
1804–1826
Jefferson Barracks
1826
Fort Scott
1842
MISSOURI
KENTUCKY
Fort Gibson
1824
Fort Smith
1827
TENNESSEE
ARKANSAS
Fort Washita
1842
Fort Towson
1824
Mississippi R.
MISSISSIPPI
ALABAMA
LOUISIANA
TEXAS
Fort Jesup
1822
New Orleans

Map 2

1,330 graduates, of whom 523 were in the Army when the war began. About another five hundred graduates returned to the Army from civilian life serving primarily with the U.S. Volunteers. Although none of the Regular Army generals and few of its field grade officers (major, lieutenant colonel, colonel) had been educated at West Point, academy graduates dominated the Regular Army's junior officer ranks, creating a professional atmosphere at the unit level. The situation was reversed for the volunteers. Unlike the regulars, the U.S. Volunteers had a large number of field grade officers who had graduated from West Point, left the Army, and returned to uniform for the war. One of the most notable, Col. Jefferson Davis, who commanded the 1st Mississippi Rifles, went on to become a U.S. senator, secretary of war, and eventually the president of the Confederate States of America. However, the volunteers had few capable company grade officers (lieutenants and captains). Most of them were either appointed by governors for political reasons or elected by the enlisted men they recruited and came to their posts with little or no military experience.

At the end of the Mexican War, Congress disbanded the volunteers and reduced the Regular Army to an authorized strength of 865 officers and 8,940 enlisted men. The government eliminated the infantry regiments created during the war but retained the Regiment of Mounted Riflemen and the two companies added to each artillery regiment. Whether they remained in service or returned to civilian life, many of the men who had fought in the Mexican War—Lee, Grant, Bragg, Hunt, Henry Halleck, Joseph Hooker, George B. McClellan, George Sykes, William T. Sherman, P. T. G. Beauregard, A. P. Hill, Thomas J. Jackson, James Longstreet, and George Pickett to name a few—would re-emerge on the national stage to lead Union and Confederate armies during the American Civil War.

Bvt. 2Lt. Ulysses S. Grant
(James Bultema Collection)

The Army Moves West

The dramatic increase in the size of the United States that resulted from the Treaty of Guadalupe Hidalgo, an 1846 treaty with Britain that added over 285,000 square miles in what is now Oregon and Washington, and an 1853 agreement with Mexico that gained about another 30,000 square miles in southern Arizona and New Mexico, required the Army to shift its resources west. Prior to the Mexican War, the Army had fifty-six military posts in the United States, only twelve of which were west of the Mississippi River. When thousands of settlers headed for California and Oregon after the Mexican War, the Army established new forts to protect the migration. By 1860, the Army had seventy-six forts west of the Mississippi River alone. Some remained in use for only a few years while others remained garrisoned for decades.

For administrative purposes, the Army divided the nation into military departments. These changed over time but eventually numbered seven by 1860. (See *Map 3*.) A brigadier general or a colonel commanded each department. Because colonels who commanded departments were also regimental commanders, such individuals relied on a lieutenant colonel to run their regiment.

By 1860 the Department of the East comprised all of the country east of the Mississippi River. The Department of the West stretched from the Mississippi River to the Rocky Mountains and from the Canadian border to the Rio Grande. The department had its headquarters in St. Louis, Missouri. The northwest corner of the country constituted the Department of Oregon which included Oregon and Washington Territory. Its headquarters was at Fort Vancouver, Washington. The Department of California included the state of California as well as portions of Utah and New Mexico Territories. The Presidio of San Francisco served as its headquarters. The Departments of Utah and New Mexico lay east of California and consisted of the territories of the same names except for those portions assigned to the Department of California. The Department of New Mexico had its headquarters at Santa Fe, while Camp Floyd served as the headquarters of the Department of Utah. Finally, the Department of Texas embraced the state of that name with its headquarters at San Antonio.

The reduction of the Army at the end of the war left the military establishment with fifteen regiments—eight infantry, four artillery, and three mounted—to garrison the seven departments. The infantry and mounted regiments moved from Mexico to western posts while the artillery regiments deployed to fortifications along the Atlantic seaboard. The 1st, 2d, 3d, and 4th Infantry initially went to Louisiana, the 5th deployed to Arkansas and Indian Territory, while the 6th, 7th, and 8th Infantry traveled to Jefferson Barracks, Missouri. The 1st and 2d Artillery went to Governor's Island, New York, while the 3d and 4th assembled at Fort Monroe, Virginia. The Regiment of Mounted Riflemen went to Fort Leavenworth and the two dragoon regiments remained in the southwest to patrol the Mexican border along the Rio Grande. From these locations the regiments further broke down and dispersed in small detachments to bring security to America's long coast and expansive frontier.

Contrary to the Hollywood image, very few of the so-called forts in the American West were actually fortified. Western forts were designed primarily as shelter for small communities of officers, enlistedmen, family members, andciviliancampfollowers, rather than as defensive structures. They were usually built by the soldiers who established and occupied them, and the construction varied widely, depending on the skills of the workmen and the availability of building materials. The soldier-builders preferred lumber, but brick, stone, adobe, and in some cases brush, had to be used depending on the location.

One of the oldest posts west of the Mississippi River was Fort Leavenworth, initially established by Col. Henry Leavenworth, 3d Infantry, in 1827. Its location near the start of both the Santa Fe

The second Fort Kearny in 1849
(Nebraska State Historical Society)

C A N
WASHINGTON TERRITORY
Fort Vancouver
UNORGANIZED TERRITORY
DEPARTMENT OF OREGON
OREGON
NEBRASKA TERRITORY
UTAH TERRITORY
Camp Floyd
DEPARTMENT OF
Presidio of San Francisco
DEPARTMENT OF UTAH
KANSAS TERRITORY
CALIFORNIA
DEPARTMENT OF CALIFORNIA
Santa Fe
NEW MEXICO TERRITORY
DEPARTMENT OF NEW MEXICO
DEPARTMENT OF T
TEXAS
Rio Grande
MEXICO
San Antonio
MILITARY DEPARTMENTS
1860
Military Department Boundary
Department Headquarters
0 100 200 300 400
Miles

Map 3

A D A
MINNESOTA
WISCONSIN
MICHIGAN
MAINE
VERMONT
NEW HAMPSHIRE
MASSACHUSETTS
Troy
NEW YORK
CONNECTICUT
RHODE ISLAND
IOWA
THE WEST
PENNSYLVANIA
NEW JERSEY
INDIANA
OHIO
MARYLAND
DELAWARE
ILLINOIS
WASHINGTON
VIRGINIA
Fort Monroe
St. Louis
DEPARTMENT OF THE EAST
MISSOURI
KENTUCKY
NORTH CAROLINA
TENNESSEE
N
ORY
ARKANSAS
SOUTH CAROLINA
MISSISSIPPI
GEORGIA
ALABAMA
EXAS
LOUISIANA
FLORIDA
Gulf of Mexico

and Oregon trails made it the gateway to the West for thousands of settlers in the 1850s. Fort Leavenworth also served as a staging area for a number of expeditions including campaigns against the Sioux in 1855 and the Cheyenne in 1857.

In June 1849, the Army purchased a privately owned trading post along the Laramie River that had been there since 1834. The Army acquired the property in response to a congressional mandate that it establish military stations along the Oregon Trail. Unlike most forts in the West, the trading post had been fortified with walls. William "Buffalo Bill" Cody described the original structure as having "walls twenty feet high and four feet thick, encompassing an area two hundred and fifty feet long bytwohundred wide." "No military frontier post in the United States was as beautiful as Fort Laramie," Buffalo Bill went on to write. It "was an oasis in the desert." Its first garrison, two companies of the Regiment of Mounted Riflemen and a company of the 6th Infantry, lived in the existing buildings until new ones could be built outside the trading post walls in the tradition of the Army's other frontier forts. Fort Laramie became a popular stopping point for travelers along the Oregon Trail. The post register recorded that almost ten thousand wagons passed through in 1850, and in 1852 an estimated forty thousand people traveled by Fort Laramie. The post also played an important role as a staging area for expeditions against hostile Native Americans in the Great Plains. (*See Map 4.*)

Further south, in New Mexico Territory, Lt. Col. Edwin V. Sumner, 1st Dragoons, established Fort Union in 1851 to protect the Santa Fe Trail and local settlements from Jicarilla Apache and Ute Indians. The new fort soon became a mainstay of the southwestern defense system in New Mexico Territory, providing both operational and logistical functions. Troops based at Fort Union conducted operations against American Indian tribes, patrolled the Santa Fe Trail, and escorted stage-coaches carrying mail.

In addition to providing protection along the major migration routes, the Army built a series of forts in Texas to protect travelers along the road from San Antonio to El Paso, to control hostile Apache and Comanche Indians in the area, and to guard the border. In 1854, Lt. Col. Washington Seawell, 8th Infantry, established Fort Davis, named for Jefferson Davis, the secretary of war at the time. Six companies of the 8th Infantry began building the post, using local materials to construct what were known as jacal-type buildings in which cottonwood strips were tied together around a frame and filled with mud, clay and grass, providing rudimentary shelter from the elements.

The American West encompassed a huge expanse of land, and the Army was hard-pressed to maintain a presence everywhere soldiers were needed. Most of the Army's frontier forts were garrisoned by only one or two companies. The 6th Infantry's experience typified how the Army dispersed to cover the frontier. The regiment assembled briefly at Jefferson Barracks, Missouri, after returning from Mexico in July 1848, but by 31 December the regimental headquarters had moved to St. Louis. Company D remained at Jefferson Barracks, but Companies A, E, and K were at Fort Snelling, Minnesota; B and F at Fort Crawford, Wisconsin Territory; C at Fort Atkinson, Kansas Territory; G and I at Fort Leavenworth; and H at Fort Scott, Kansas.

Most of the Army's regiments spent considerable time in the field, and companies moved frequently to meet changing requirements. According to its regimental history, the 2d Cavalry spent most of "its time under canvas, with a certainty of constant scouting and a change of station at least once a year." Long marches were common. While campaigning in 1851, Company B, 1st Dragoons, rode 2,240 miles across the central Great Plains. In 1856, the 3d Infantry covered 500 miles in New Mexico Territory searching for Gila and Mogollon Apaches. During three months in 1858, the 6th Infantry marched over one thousand miles from Fort Bridger, Utah Territory, to Benicia Barracks, near San Francisco, California. Capt. Winfield Scott Hancock, a future Civil War general who participated in the march, observed that "the men bore the fatigue well, and their general conduct is favorably represented by the officers." Two years later, Battery B, 4th Artillery, traveled over two thousand miles across "a barren and desert country" to protect mail routes from hostile Indians in Utah Territory.

The Army and Native Americans

The Army had been interacting with American Indians long before the Mexican War. By the end of the War of 1812, the United States had conquered most of the tribes living east of the Mississippi River. One important exception was Florida's Seminole Indians. Prior to the Mexican War the United States had fought two costly wars in Florida that had resulted in the removal of the

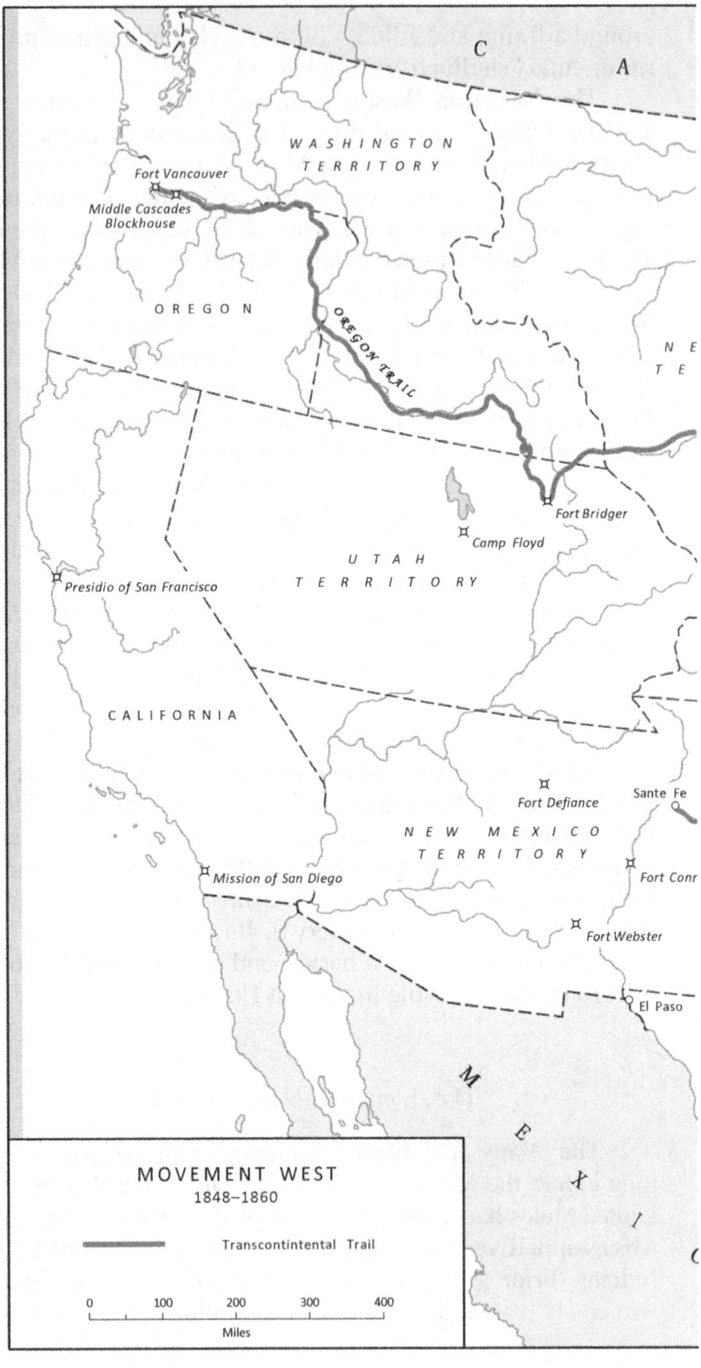
C
A
WASHINGTON
TERRITORY
Fort Vancouver
Middle Cascades
Blockhouse
OREGON
OREGON TRAIL
N E
T E
Fort Bridger
Camp Floyd
UTAH
TERRITORY
Presidio of San Francisco
CALIFORNIA
Fort Defiance
Sante Fe
NEW MEXICO
TERRITORY
Mission of San Diego
Fort Conr
Fort Webster
El Paso
M
E
X
I
MOVEMENT WEST
1848–1860
Transcontintental Trail
0
100
200
300
400
Miles

Map 4

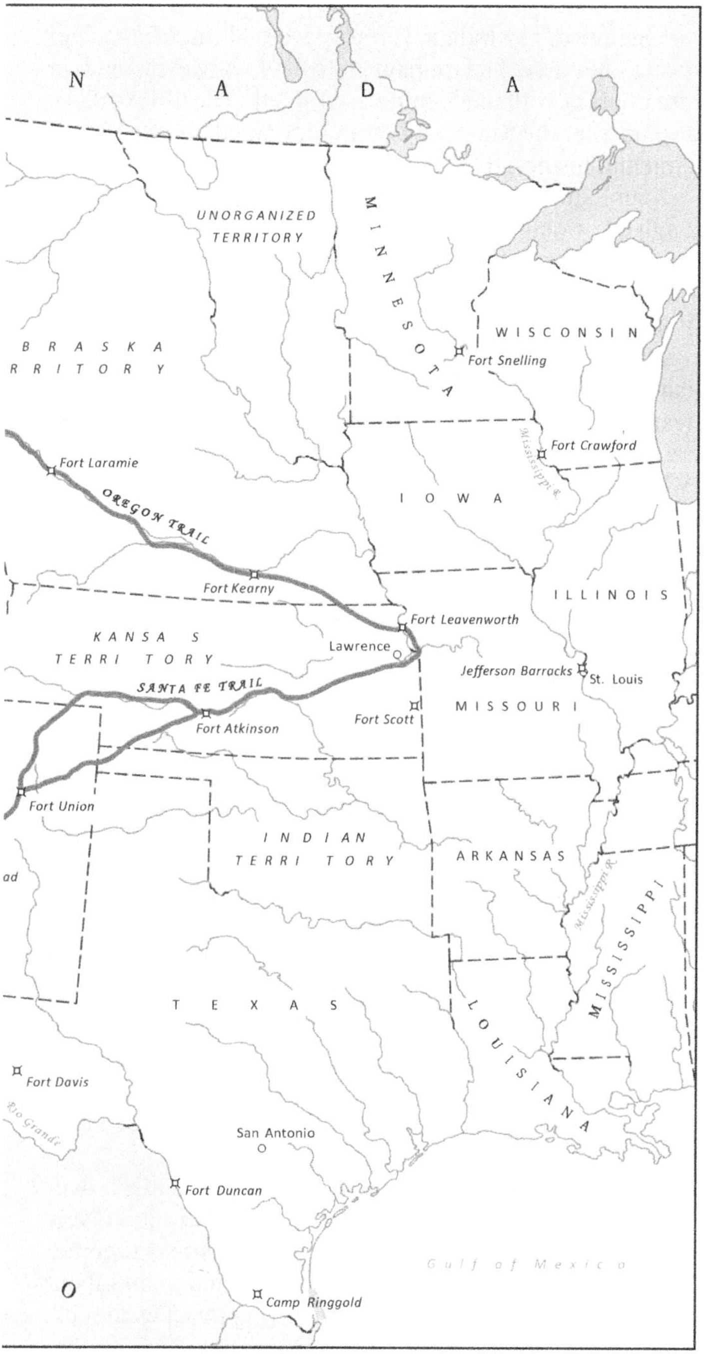

N A D A
UNORGANIZED TERRITORY
MINNESOTA
WISCONSIN
Fort Snelling
B R A S K A
R R I T O R Y
Fort Crawford
Mississippi R.
Fort Laramie
OREGON TRAIL
I O W A
Fort Kearny
ILLINOIS
Fort Leavenworth
K A N S A S
T E R R I T O R Y
Lawrence
Jefferson Barracks
St. Louis
SANTA FE TRAIL
Fort Atkinson
Fort Scott
M I S S O U R I
Fort Union
I N D I A N
T E R R I T O R Y
ARKANSAS
ad
Mississippi R.
MISSISSIPPI
T E X A S
LOUISIANA
Fort Davis
Rio Grande
San Antonio
Fort Duncan
Gulf of Mexico
O
Camp Ringgold

majority of Seminoles to Indian Territory west of the Mississippi River. Several hundred had remained behind, however, and in 1855 a third conflict with the Seminoles erupted. The 4th Artillery spent the years after the Mexican War in Florida where, according to the regimental history, it "was very actively engaged in hunting Indians . . . and suffered great hardships in some of its expeditions through the swamp." The fighting in Florida finally ended in 1858 when all but a handful of American Indians agreed to move to Indian Territory. The few Seminoles who refused to relocate withdrew into the Florida Everglades where they refrained from further confrontations with the white population.

Indian-White conflict was rare east of the Mississippi after the Mexican War; however, the situation was different west of the river. Here Indians resented and resisted the growing encroachment of white settlers on their lands, be they in the Great Plains, the mountains of the Pacific Northwest, or the Southwestern deserts. The Army's primary mission in the West was to protect American civilians, particularly along the major migration routes such as the Oregon and Santa Fe trails. In an effort to make the Oregon Trail safer, in the summer of 1851 the United States signed a treaty at Fort Laramie that called for peace with the Indian tribes that populated the northern Great Plains. Some ten thousand Indians representing the Cheyenne, Arapaho, Crow, Shoshone, Assiniboine, Mandan, Hidatsa, and Arikara assembled for the largest council in recorded history. Under the terms the Native Americans agreed to allow roads and forts to be built in their territories while the U.S. government promised to protect the Indians from hostile settlers and to provide an annual payment of $50,000 worth of goods for fifty years. Although Congress later cut the annuity period to ten years, and some tribes received nothing, the treaty did bring a brief period of respite along the Oregon Trail. Two years later the government met with Comanche, Kiowa, and other tribes of the Southern Plains at Fort Atkinson, Kansas, to make the Santa Fe Trail safer. The treaty that resulted resembled the one signed at Fort Laramie and helped to keep the peace in the Southwest for a few years as well.

The Army's relationship with individual tribes changed over time. An example is the Army's relations with the Navajo in New Mexico Territory. In 1853, Col. Joseph K. F. Mansfield reported that at Fort Defiance "the Indians are freely admitted in and about the fort during the day and are friendly and bring peaches to sell."

But as more settlers moved into the area, the Navajo grew resentful. In April 1860 some one thousand Navajo warriors attacked and briefly occupied parts of the fort before the garrison drove them out. The attack marked one of the few times Native Americans in the West assaulted an Army fort.

Far more common than assaults on a major post were fights in the open countryside. Most of these engagements were small affairs with few casualties on either side. Nevertheless, a seemingly small incident could provoke a major fight. In August 1854, a settler on a westbound wagon train registered a complaint at Fort Laramie that a Sioux Indian had butchered a lame ox. Second Lt. John L. Grattan, just one year out of West Point, led a force of thirty men from Company G, 6th Infantry, along with two artillery pieces and their crews to apprehend the culprit. Although advised to be cautious, the young lieutenant led his force directly into an Indian camp about eight miles from the fort and demanded that the guilty party surrender. A tense standoff ensued that soon led to gunfire. While it is not clear who fired the first shot, Indian

Fort Defiance, New Mexico, *by Seth Eastman*
(Architect of the Capitol)

warriors quickly overwhelmed Grattan and his men. A single wounded soldier made it back to Fort Laramie to report the incident before he died of his wounds.

In August 1855 the Army launched a concerted effort to find the Indians responsible for the "Grattan Massacre." Col. William S. Harney, commander of the 2d Dragoons, organized a force of more than six hundred men that included companies from a several different regiments. Intending to strike "a decisive blow against any one of the hostile bands of Sioux," Harney arrived on the North Platte River on 2 September. Six miles away about four hundred Brule Lakotas had established a village. About one hundred twenty of the Indians were of fighting age. Harney knew that the village leader, Little Thunder, had been one of the men implicated in the Grattan killings. Planning to attack quickly, he sent four mounted companies across the North Platte River, where they would be north of the village, to block a retreat by the Brules. The infantry moved into position to attack from the south. When the mounted units heard gunfire they were to join the attack. As the infantry moved forward they encountered a small party of Brules on horseback. Anticipating trouble, Little Thunder ordered the villagers to strike their tepees and move north. He then rode to meet Harney under a flag of truce. Harney demanded that Little Thunder turn over the men responsible for the Grattan Massacre and other disturbances. Little Thunder refused. Harney ordered his troops to advance. As the Indians tried to flee they were caught between the infantry and the mounted troops. The toll amounted to eighty-six Brules killed and five wounded along with about seventy women and children taken prisoner. Seven enlisted soldiers died and five were wounded. Little Thunder escaped. A search of what remained of the village revealed army clothing that presumably belonged to Grattan's men and papers from mail robberies.

One of the more active theaters in the decade before the Civil War was the Pacific Northwest, where the Army attempted to pacify the indigenous inhabitants and force them onto reservations. Conflict inevitably resulted. Two examples from 1856 illustrate the nature of these fights. In March 1856, about one hundred Yakima, Klickitat, and Chinook warriors surrounded a blockhouse manned by eight soldiers at Middle Cascades, Washington Territory. Upon learning of the siege, 2d Lt. Philip H. Sheridan gathered about forty infantrymen from the 4th Infantry at Fort

Vancouver, loaded them aboard a steamboat, and headed up the Columbia River. At about the same time, Col. George Wright, 9th Infantry, sent some two hundred fifty soldiers down the river on two steamboats to join Sheridan. The combined force lifted the siege, but the Yakimas and Klickitats escaped. Colonel Wright tried nine Chinook tribal leaders at a military tribunal and hanged eight of them. Both Sheridan and Wright would become generals a few years later during the Civil War.

At about the same time as the siege, four companies from the 4th Infantry, two from the 3d Artillery, and one from the 1st Dragoons under Maj. Robert C. Buchanan set out in Oregon Territory to force the American Indians of the Rogue River area to move onto the Coast Reservation. After two months of marching through damp, cold mountains, Buchanan invited the Indians to a peace conference in late May. The Indians promised to surrender, but when a detachment of about eighty soldiers and a howitzer assembled at the designated meeting place at Big Meadows, Oregon Territory, a pair of Native women warned them that a Rogue Indian chief known as Old John planned to attack them the next day. After sending a courier for reinforcements, the men dug in and with the help of the howitzer repulsed Old John and his 200 warriors on 27 May 1856. By the following day, however, the situation was desperate, with a third of the soldiers killed or wounded, their water gone, and their ammunition running low. The Native Americans were massing for a final assault when reinforcements in the guise of a company of the 4th Infantry led by Capt. Christopher C. Auger suddenly appeared behind the Indians. The defenders rallied and charged the warriors who retreated, taking their dead and wounded with them. The fight at Big Meadow effectively ended the Rogue River War, with some twelve hundred Indians surrendering to reservations by the end of June. Like Sheridan and Wright, Buchanan and Augur were destined to become Union generals during the Civil War.

The Army and the Settlers

Although the Army's mission in the West centered on protecting settlers from American Indians, on occasion soldiers were called on to protect Indians from whites, be they trespassers, unlawful settlers and prospectors, unscrupulous merchants, or hateful vigilantes. A number of Army officers had little regard for some of the civilians they encountered on the frontier. They were struck by the low regard frontiersmen had for human life. Lt. Richard Johnson wrote from Fort Davis that "people are ignorant, destitute of any refinement, and have no respect for the laws." Another officer described Texas as "a country where little is known of, less cared for, the laws of God and man." Most Army officers wanted to resolve problems between Native Americans and settlers without resorting to violence, a view often not shared by white civilians living on the frontier. Capt. Thomas W. Sweeney expressed his opinion that "all our *Indian wars*, with very few exceptions, are brought on either by our *frontier settlers* . . . or the traders in the Indian Country, who as a class, are an unmitigated set of *scoundrels*" (emphasis original).

Trying to keep the peace between whites and American Indians was a difficult and often thankless job, but soldiers were also called on to keep the peace among whites themselves. On two occasions, the government deployed substantial forces for such a purpose. The first of these events occurred in Kansas, the second in Utah.

In 1854, the question of whether slavery should be permitted to spread westward provoked civil disturbances in Kansas. The Missouri Compromise of 1820 had prohibited slavery north of 36°30′, but during its efforts to organize the Kansas-Nebraska Territory, Congress passed the Kansas-Nebraska Act that repealed the Missouri Compromise and left the slavery question up to local authorities to resolve. The inevitable outcome of this decision was a fierce and sometimes violent debate in Kansas between proslavery and antislavery settlers. By 1855, two rival governments existed in the territory, one proslavery, the other antislavery. The situation came to a head in May 1856, when proslavery forces sacked Lawrence, an antislavery, or free-soil, town. Several days later the noted abolitionist, John Brown, and six of his followers retaliated by executing five proslavery men. The two events ignited numerous encounters between armed bands from both sides.

With lawlessness becoming endemic, the territorial governor, Wilson E. Shannon, requested assistance from the federal government. Colonel Sumner, commander of the 1st Cavalry and Fort Leavenworth, sent mounted troopers to patrol the roads between Lawrence and the slave state of Missouri. Nevertheless, for the rest of the summer chaos reigned in Kansas as free-soil and proslavery forces terrorized each other. In the summer of 1857, a force of antislavery partisans known as Jayhawkers entered the fray, skirmishing against proslavery guerrillas. Although Army troops tried to bring them to heel, the Jayhawkers repeatedly managed to elude the troops and disappear into the prairie lands. Not until 1858 did Capt. Nathaniel Lyon, who by then headed the pacification program, manage to bring calm to the territory. In August, Governor John W. Denver, confident that peace had finally returned to the state, dismissed the regulars and replaced them with mounted volunteers.

Concurrent with the events in "Bleeding Kansas," instability emerged in Utah Territory. The trouble in Utah had its roots in 1847 when Brigham Young led his followers, the Church of Jesus Christ of Latter-day Saints, also known as Mormons, to the valley of the Great Salt Lake where they decided to settle. At the time the area belonged to Mexico, and the Mormons, who had been persecuted in the East, believed that in Utah they would be free of American laws and prejudices. However, in 1848 the United States acquired the area from Mexico as part of the Treaty of Guadalupe Hidalgo. Mormon leaders then wrote a constitution and appealed to Congress to recognize their land as the State of Deseret where they could control local laws. The U.S. government instead imposed territorial status on Utah. Young became the governor, but relations were tense as many Americans were suspicious of the Mormon faith and frowned on the Mormons' practice of polygamy. Young consolidated his power by appointing Mormon bishops to the probate courts, establishing what many saw as a theocracy in the territory.

Trouble between the Mormons and the federal government continued to fester with periodic conflicts between Army troops and Mormon militia. During the 1856 presidential election polygamy became an issue and shortly after his inauguration in March 1857, President James Buchanan ordered the Army into the Great Salt Lake Valley. Troops from the 4th, 5th, and 10th Infantry, the 2d Dragoons, and the 4th Artillery assembled at Fort Leavenworth in preparation for the operation. In July the Utah Expedition left the fort with 500 men, 97 wagons, and 600 hundred animals under the command of Col. Edmund B. Alexander, the elderly commander of the 10th Infantry. Upon learning of the expedition, Young prepared for war while Mormon rangers harassed the federal column.

Faced with the probability of armed conflict with the Mormons and an unseasonable blizzard, Alexander retreated. In the meantime, the War Department ordered Col. Albert S. Johnston, commander of the 2d Cavalry and future Confederate general, to take command of the faltering expedition. Arriving on the scene from Texas in November 1857, he led the weary troops into winter quarters at Fort Bridger, Utah, and began planning for a spring campaign. Shortly thereafter the newly appointed territorial governor, Alfred E. Cumming, arrived at Fort Bridger. Johnston and Cumming believed that the situation had to be settled by force, but in June 1858, peace commissioners sent by President Buchanan arrived in Utah and negotiated a settlement without bloodshed. The so-called Mormon War never happened. The Army occupied several forts in Utah Territory until 1861, when it sent the men East at the beginning of the Civil War.

Army Life on the Frontier

The years before the Civil War found the Army largely isolated from the rest of the nation. The small garrisons in the West where most of the soldiers were stationed were lonely outposts. Officers and noncommissioned officers who were assigned to the War Department in Washington, D.C., or to one of the depots or arsenals in the East, as well the artillerymen who garrisoned coastal fortifications, lived reasonably pleasant lives. Their fellow soldiers stationed on the frontier regularly endured hardship, disease, poor rations, and less than desirable living quarters.

Soldiers in the nineteenth-century Army generally spent their entire military career in the same regiment, whether serving for a single enlistment or a lifetime. Many who served on the frontier prior to the Civil War were from Northern cities. They included young boys and men seeking adventure, men escaping problems at home, and European immigrants. As an example of the diversity in the ranks, when the 10th Infantry began recruiting in 1855, of the first 500 men who enlisted in the new regiment, 66 were born in New England, 149 were from the West and Midwest, and 285 were foreign-born. The regimental history notes that 55 percent of the recruits deserted before completing their enlistments.

Other than the occasional encounter with hostile Native Americans, life on the frontier could be tedious. Capt. John W. Phelps, a company commander serving at Camp Floyd, wrote in 1859: "I am suffocating, physically, morally and intellectually—in every way. I am fairly gasping for fresh outside air, and feel as an officer said the other day, like begging to be taken out and hung for the sake of variety." Enlisted soldiers likewise found garrison life dull. Eugene Bandel, a company first sergeant in the 6th Infantry, wrote to his parents in June 1859 that "there is very little to vary the monotony as the weeks go by." First Sgt. Percival Lowe with the 1st Dragoons in New Mexico, on the other hand, recalled how his unit battled boredom by forming a thespian society that performed weekly for the officers and their wives.

Frontier forts varied in their amenities. In 1853 Colonel Mansfield toured New Mexico Territory and reported that the buildings at Fort Union were "as good as at any post and there seems to be enough of them to satisfy the demands of the service." At Fort Webster, however, he found the construction "quite indifferent and not sufficient for the command," while at Fort Conrad, "The quarters of both officers and soldiers are falling to pieces." The following year he inspected posts in California and was once again largely unimpressed. He found the barracks at the Presidio of San Francisco to be "miserable adobe buildings," while the living quarters at the Mission of San Diego were "worthless" for the men and "quite indifferent and not suitable" for the officers.

Despite shortcomings in their quarters, the occupants generally made the best of their situation. When Teresa Griffin Viele, wife of Lt. Egbert L. Viele, 1st Infantry, got her first look at Camp Ringgold, Texas, in 1851, she thought the buildings "all reminded me of the house of the foolish man, 'who builds his foundation upon the sand,' all being in the same plight. There were no signs of vegetation around; not even a blade of grass." In spite of that dismal first impression, she looked back on the "year passed in this isolated spot as one full of pleasant recollections." Compared with facing the elements when out in the field, Lt. Edward L. Hartz reflected that his quarters at Fort Davis were "decidedly pleasant when returning fagged out from constant traveling, bivouacking and hard feeding. They offer you a comfortable bed, a roof to shelter, and the enticements of a tolerably well spread table."

Expansion and Modernization

Even as its soldiers engaged in their long-standing duties as a frontier constabulary, the Army endeavored to modernize. When the Mexican War had ended in 1848 the Army reverted to a peacetime strength of fewer than ten thousand soldiers, essentially the same strength it had been authorized in 1815 at the conclusion of the War of 1812. However, as demands of frontier service increased, Congress raised the number of privates in each company stationed in the West to 74, making the authorized strength of the Army 12,927 officers and men. Then, in 1855, Secretary Davis persuaded Congress to add four regiments to the Army, two each of infantry and cavalry. The 9th Infantry formed at Fort Monroe and the 10th Infantry assembled at Carlisle Barracks, Pennsylvania, after which both regiments moved to the Western frontier. The 1st and 2d Cavalry organized at Jefferson Barracks. The 1st Cavalry then moved to Fort Leavenworth while the 2d went to Texas. With the two new cavalry regiments, the Army now had three distinct kinds of mounted troops—dragoons, mounted riflemen, and cavalry. Each carried different weapons and wore

The American Soldier, 1855, *by H. Charles McBarron*
(U.S. Army Art Collection)

different colors to denote their branch—orange for the dragoons, green for the mounted riflemen, and yellow for the cavalry. Inpractice, little distinguished how they operated in the field. As a result of the augmentations of the 1850s, the Army grew to over sixteen thousand men by the end of the decade.

About the same time that the Army increased its force structure, it upgraded its standard infantry weapon when it began replacing smoothbore muskets with rifled ones. Rifled weapons had been around for many years. Rifling—spiral grooves cut in the bore of a gun barrel—caused projectiles to spin rapidly, stabilizing them in flight and giving them greater accuracy and longer range. However, rifles required a bullet slightly larger than the bore to catch the grooves, which meant that soldiers had to pound the round down the barrel, a difficult task in combat that significantly reduced the rate of fire. Well-trained infantrymen using a smoothbore could load and fire three or four times faster than those with a rifled musket.

By the middle of the nineteenth century, technological developments made possible an accurate, dependable muzzle-loading rifle that could be loaded as quickly as the smoothbore musket. In the late 1840s a Frenchman, Capt. Claude E. Minié, invented an oblong bullet with a hollow iron cup at the base that fit easily into the barrel when loading but which expanded upon being fired so as to catch the rifling. European armies quickly adopted the new minié round. The U.S. Army tested the concept in 1854

New rifle-musket ball, caliber .58," 1855
(Smithsonian)

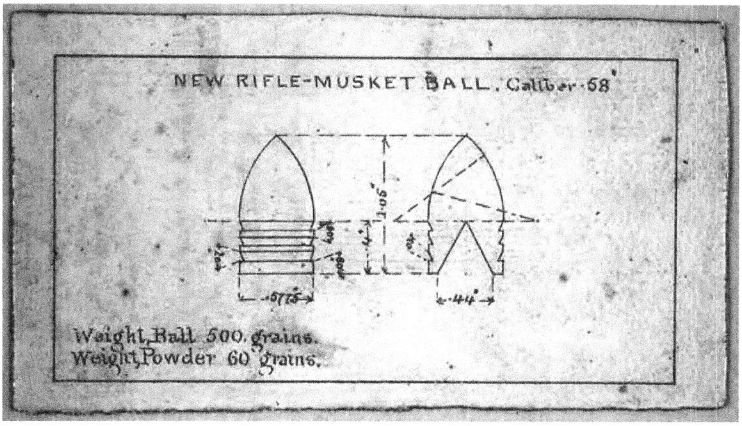

and pronounced it superior to the venerable smoothbore. The next year, the federal armories at Springfield, Massachusetts, and Harper's Ferry, Virginia, began converting old smoothbores into rifles by adding rifling to the bores. They also began producing a new weapon, the Springfield Model 1855 rifle-musket, manufacturing four thousand by the end of 1858.

The Army also explored the possibility of developing a breechloading firearm that could be reloaded more rapidly than a muzzleloading weapon. In 1854, Congress appropriated $90,000 to test and purchase a breechloader. A series of boards examined a variety of proposals over the next four years, but could not agree on a model to replace the muzzleloaders for the infantry. Fielding a repeating rifle offering even greater range, accuracy, and rate of fire than any muzzleloader or breechloader also proved difficult. The Colt revolving rifle model 1855, issued in limited quantities on a trial basis, frequently burned soldiers' hands when the paper cartridges accidentally ignited in the cylinder. George W. Morse corrected that problem in 1858 when he developed metallic cartridges to replace the paper ones. Two years later he began converting muzzleloaders to breechloaders at the Harper's Ferry Arsenal. That project ended in 1861 when Morse defected to the South and Confederate troops captured Harper's Ferry. Meanwhile, by 1860 Northern inventers had developed magazine-fed repeaters—such as the Spencer and Henry rifles. Therefore, by the outbreak of the Civil War America had the technology to produce breechloaders and repeaters. The U.S. Army's Ordnance Department, however, did not want to unduly complicate the manufacture and distribution of rifles and ammunition. Consequently, the muzzle-loading rifled musket remained the standard infantry weapon for both sides during the Civil War, with breechloaders and repeating rifles confined mostly to cavalry and specialist units.

The advent of rifled firearms prompted an interest in marksmanship training, something that had rarely been accomplished previously. Brig. Gen. William S. Harney, commander of the 1856 expedition against the Sioux, ordered daily target practice to ensure his soldiers would be familiar with their new weapons. Sergeant Bandel of the 6th Infantry, part of Harney's command for the campaign, recalled using "seventy-five cartridges in one day alone" soon after he received his rifled musket. In March 1856, the 10th Infantry also instituted marksmanship practice. Targets ranged from two hundred to seven hundred yards. According to the regimental history, "All shots were recorded and the men classified according to ability." The 3d Artillery regimental history noted that "the superiority of the rifle was at once strikingly manifest." In 1858 the Army codified these efforts by introducing a new target practice manual. Nevertheless, by the time the Civil War erupted, marksmanship training was still in its infancy, and in the rush to raise mass armies at the outbreak of the war, such training often fell by the wayside.

There were also improvements in artillery between the Mexican War and the Civil War. In 1857 the U.S. Army adopted the "Napoleon" gun-howitzer. The weapon was named after the French emperor Napoleon III who had introduced it to simplify his artillery by combining the functions of a howitzer and a gun (or cannon) into the same weapon. The smoothbore, bronze weapon, classified as a 12-pounder because it fired a solid shot projectile weighing twelve pounds, first demonstrated its effectiveness in the Crimean War of 1853–1856. The versatile Napoleon gun-howitzer could fire solid shot, explosive shell, spherical caseshot (shrapnel), and canister. To simplify the supply system, the ordnance department designed standardized ammunition chests that included a variety of the types of projectiles the gun could fire.

The Army worked on other aspects of artillery weapons as well. In the 1840s, 1st Lt. Thomas J. Rodman developed a method of casting iron guns thatmade them stronger during firing. Inthe mid-1850s, the Army began to experiment with rifled artillery at Fort Monroe, and in

Artillerymen of Battery C, 3d U.S. Artillery, at drill on Fort Vancouver, Washington Territory, 1860. (National Archives)

1860 a board of artillery and ordnance officers recommended that at least 50 percent of the Army's guns at forts and arsenals be converted into rifles. The idea did not prove effective because adding the groves to existing bronze cannons weakened the barrels. Iron proved to be the answer. Rifled guns, made of cast iron, began to be manufactured in a variety of calibers during the winter of 1860–1861. Capt. Robert P. Parrott of the 3d Artillery developed a method to add a reinforcing hoop on the breech to withstand the strain of firing. The 10-pounder Parrott initially had a 2.9-inch bore, but that increased to 3 inches in later models to make them compatible with the ammunition used in the 3-inch Ordnance Rifle introduced in 1861. Napoleons, Parrotts, and Ordnance Rifles would become standard equipment during the Civil War. However, not all new weapons were equally successful. When Battery G, 4th Artillery, received mountain-howitzers in May 1855, it found that "after firing a few rounds of ammunition at practice, the materiel was so much injured by the cracking of axles, etc., that the scheme had to be abandoned."

Along with better weapons came a deeper interest in professional literature. In 1846, 1st Lt. Henry W. Halleck's *Elements of Military Art and Science* stimulated the study of military theory

Smoothbore Artillery Ammunition – Cartridges and Projectiles; and Rifled Artillery projectiles (Dover Publications, Royalty Free Clip Art)

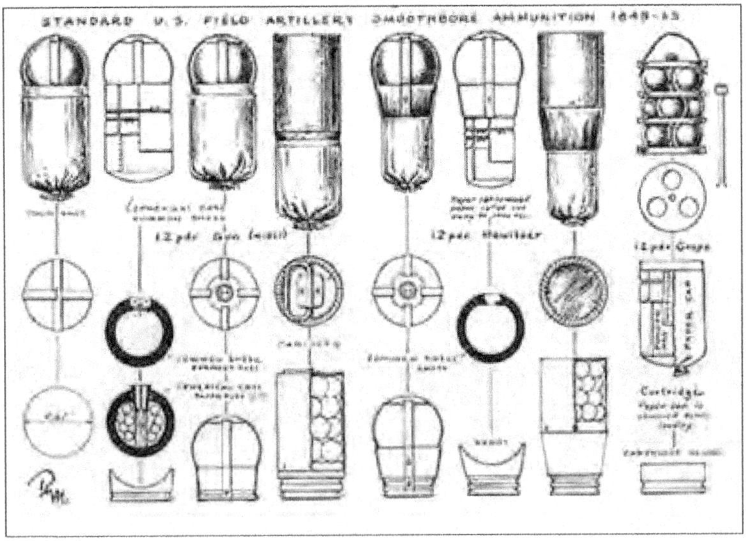

in the United States. Halleck, a graduate of West Point known as "Old Brains" because of his academic achievements, would rise to the rank of major general and command the Army for a short time during the Civil War.

Much of the new literature focused on artillery. In 1849, Maj. Alfred Mordecai's *Artillery for the United States Land Service* codified the American system of artillery. The manual contained complete drawings and descriptions of the different guns, howitzers, and mortars and their carriages that were in the Army's inventory. The artillery ranged in size from 6-pounders to 42-pounders and from 12-pounder Coehorn mortars weighing 164 pounds to the 10-inch Columbiad weighing 15,260 pounds (unmounted). In 1851 the War Department issued new artillery drill regulations. General orders issued in 1859 established a systematic course of practical and theoretical artillery instruction and authorized inspectors to examine artillery units at least once a year. Although many artillery companies were not equipped with guns in the 1850s, these measures promoted uniformity in training. Finally, in the late 1850s the War Department formed a panel to write a new artillery manual. The resulting *Instructions for Field Artillery*, published in 1861, would become the bible of Federal artillerymen during the Civil War.

15-inch Model 1861
(National Archives)

Meanwhile, after completing his work on artillery, Mordecai joined Maj. Richard Delafield and Capt. George B. McClellan in a trip to Europe to study European military institutions and the Crimean War. Their published findings, as well as those of other officers posted to Europe during the period, stimulated American study of the latest European methods, particularly those of the French. One piece of equipment that emerged from these studies was a saddle, designed by McClellan and adopted by the Army in 1859. The McClellan saddle would remain in service until the Army disbanded its horse cavalry after World War II. Less durable, but perhaps more consequential, legacies of this era were new drill and tactical forms inspired by French practice. One prominent example was *Hardee's Tactics,* published in 1855 by brevet Lt. Col. (and future Confederate general) William J. Hardee.

Europe was not the only source of inspiration for American soldiers. Based on their usage in Africa and the Middle East, in 1843 Capt. George H. Grossman proposed using camels for carrying supplies in the deserts of the American Southwest. Nothing came of the idea as he was unable to obtain funding. In 1851, Jefferson Davis, then a senator and chairman of the Senate Military Affairs Committee, tried but failed to get funds for the Army to buy thirty camels. He finally met with success as the secretary of war when he convinced Congress to appropriate $30,000 to purchase seventy-five camels and transport them from North Africa to Texas.

Camels in Texas, *by Tom Lovell*
(commissioned by the Petroleum Museum in Midlands Texas)

In April 1859, Lieutenant Hartz, acting quartermaster of the 8th Infantry, conducted a trial in the desert country of southern Texas along the Rio Grande comparing twenty-four camels and twenty-four mules to determine the "capabilities and usefulness [of camels] as a means of transportation for military purposes." From May to August Hartz led his expedition through rugged country in intense heat with the camels demonstrating that they could travel twenty miles a day carrying up to five hundred pounds with little or no water. He reported that "the superiority of the camel for military purposes in the badly watered sections of the country seems to me to be established."

The camel experiment came to an end with the beginning of the Civil War. As Regular Army troops moved east the camels were sold or released into the wild to fend for themselves in the deserts of California, Nevada, Arizona, and Texas. Some were killed and eaten by Indians. Others were shot by hunters or frontiersmen. But a few survived. A work crew for the Santa Fe railroad reported seeing a camel in 1913 near Wickenburg, Arizona. Sightings of the animals were also reported near Palm Springs, California, in 1930 and Ajo, Arizona, in 1931.

Assistant Surgeon Albert Myer in 1854 (U.S. Army Signal Center and Fort Gordon)

At about the time the Army conducted its camel experiment, another new idea took root in Texas. In 1854, Albert Myer joined the Army as an assistant surgeon. Myer had studied medicine at the University of Buffalo where he had also worked for the New York State Telegraph Company. By combining his medical education with his telegraph experience he developed a system for deaf-mutes to communicate by tapping on a person's hand or cheek. While serving at Fort Duncan, Texas, he adapted this sign language into a signal system using a single flag. Letters were transmitted by waving the flag back and forth in a set pattern for letters and numbers. The system became known as wigwag. In 1856, Myer offered his system to the War Department. The Army's Chief of Engineers, Col. Joseph G. Totten, supported the idea. Secretary Davis did not. When John B. Floyd replaced Davis as secretary in 1857, Totten reintroduced the proposal. Floyd invited Myer to appear before a board of examination in Washington. The board, headed by Lt. Col. Robert E. Lee, met in March 1859 and recommended further study of the proposed system.

Based on the board's recommendation, Myer, assisted by several other officers including 2d Lt. Edward P. Alexander, an engineer officer, began tests at Fort Monroe. The team soon communicated at distances up to fifteen miles. In late November, Myer reported that the tests exceeded expectations and suggested that the Army adopt the system and put him in charge of it. In February 1860, Myer and Alexander appeared before the Senate Committee on Military Affairs, chaired by Senator Davis, and the House Committee on Military Affairs. Based on Myer's testimony Congress included $2,000 dollars in the Army's 1861 budget and approved adding a signal officer with the rank of major to the force structure. In June 1860, the War Department appointed Myer to the position. Within a month of being appointed, Major Myer set out to New Mexico Territory where he worked in the field with Maj. Edward R. S. Canby of the 10th Infantry. The system proved effective during operations against the Navajos, prompting Canby to propose a corps of officers who would specialize in signaling, an idea Myer embraced. In May 1861, with war on the horizon, Myer reported for duty at Fort Monroe where he established a school to train personnel in his signal system. This marked the birth of the Signal Corps.

The Army and Civil Works

A young nation like the United States could hardly afford to confine its use of the talent embodied in its officer corps to purely military functions. From the first the government had relied on the Army to explore new territories, and with the U.S. Military Academy offering the only courses in engineering in the United States until the 1830s, it had logically become a major source for skilled builders, designers, and cartographers. Prior to the war with Mexico, Army engineers had helped plan roads, railroads, and canals. After the war, Army engineers continued to help build the nation.

Army topographical engineers had begun supervising the construction of lighthouses in 1834, and when the Federal government established a Lighthouse Board in 1852, three Army officers served as members. The board supervised lighthouse construction and inspection. Army engineers also served in each of the twelve lighthouse districts and designed and built a variety of lighthouses. Seminole and Mexican War veteran George G. Meade excelled at the task. As a lieutenant and captain in the 1850s Meade designed and built several lighthouses in New Jersey including those at Long Beach Island, Atlantic City, and Cape May, and in Florida at Jupiter, and in the Florida Keys. Meade would then go on to survey the Great Lakes. Another Army engineer, Maj. Hartman Bache, borrowed a design from British engineers to build the first screw-pile lighthouse in the United States at the mouth of the Delaware Bay. Twisting the screw piles into the bottom of the bay secured the structure to the seafloor. Both Meade and Bache were destined for general officer rank, with Meade rising to command the Army of the Potomac during the Civil War.

In Washington, D.C., Army engineers built aqueducts, bridges, and public buildings. In 1853, Lt. Montgomery C. Meigs,

Topographical Engineers exploring the Colorado River near Chimney Peak
(National Archives)

future Quartermaster General for the Union Army during the Civil War, received the mission of constructing a permanent water supply for the city. The project included an aqueduct with two bridges that later carried traffic as well as water pipes across the Cabin John and Rock creeks. In the years leading up to the Civil War Meigs also supervised additions to the Capitol which included the House and Senate Wings and the dome.

Prompted by the periodic floods that occurred along the lower Mississippi River, Congress appropriated $50,000 for a topographical and hydrographical survey of the Mississippi Delta in September 1850. The survey included a study of how best to develop a twenty-foot navigation channel at the mouth of the river. Capt. Andrew A. Humphreys initiated the survey. In 1857, 2d Lt. Henry L. Abbot joined him. Four years later the two men coauthored "Report upon the Physics and Hydraulics of the Mississippi River," a document that influenced the development of river engineering and the evolution of the Corps of Engineers for the next sixty years. Both men would become generals in the Union Army.

West Pointers Bide Their Time

The years between the end of the Mexican War and the start of the Civil War were ones of conflict, innovation, and industry, and those who fought, thought, and built received few rewards. Promotion was slow in the antebellum Army, and a West Point education did not make the process any faster. Between its founding in 1802 and the beginning of the Civil War in 1861 the U.S. Military Academy graduated 1,966 cadets, about half of whom served in the Mexican War. By 1861, 750 of the 1,063 graduates still living were on active service in the Army. Even though West Point had been producing officers for more than fifty years and about 75 percent of the Army's 1,098 officers in 1861 were graduates, few of them served in senior leadership positions. None of the Army's four general officers at the start of the Civil War were West Point graduates. Indeed, three of the four were veterans of the War of 1812. Of the nineteen regimental colonels, only six had graduated from the U.S. Military Academy.

The promotion system had little or no room for flexibility. It all depended on an officer's date of rank. Officers were promoted within the regiment as openings became available, usually when an officer left the Army or died. With no retirement program, officers tended to remain on active service as long as possible, regardless of age or physical infirmities. Officers spent many years as lieutenants and captains waiting for a vacancy to open up ahead of them. In the years before the Civil War it took an average of forty-four years for an officer to reach the rank of colonel in the artillery. The infantry fared a bit better, taking an average of only thirty-seven years of service to attain that grade. In the mounted arms, it took just twenty-one years of service to be promoted to colonel, because when Congress added two cavalry regiments to the Regular Army in 1855, it created vacancies for field grade officers who were selected from the three existing mounted regiments in the Army, thereby lowering the average promotion time.

Officers could gain rank outside the rigid seniority system by means of a brevet. The brevet, usually awarded for gallantry in battle, entitled an officer to wear the insignia of a higher rank, although it did not authorize higher pay. In the Mexican War, many officers received brevet ranks that they continued to hold after the war while they waited to be promoted to permanent ranks within the regimental structure. Lee, for example, a captain in the engineers, received three brevets for distinguished service in Mexico, allowing him to wear the rank of colonel for many years before he was formally promoted to that rank in 1861. Although appointed superintendent of the U.S. Military Academy at West Point in 1852, his permanent rank remained a captain ofengineers until 1855 when Secretary Davis appointed him the lieutenant colonel of the newly organized 2d Cavalry. Six years later Lee received a promotion to colonel in the 1st Cavalry,

Major Joseph Hooker
(Courtesy of William Schultz Collection)

but he resigned shortly thereafter to take command of the state military forces of Virginia.

Not all West Point graduates opted for Army life—313 of those still living in 1861 had resigned to try their hand at civilian careers with mixed success. Some thrived as civilians; others faltered. Two men who faltered were Ulysses S. Grant and William T. Sherman. After receiving brevet promotions to the rank of captain during the Mexican War, Sherman resigned in September 1853 and Grant left in July 1854. Sherman tried banking and law before becoming the superintendent of the Louisiana State Seminary of Learning and Military Academy in Pineville, Louisiana. Grant tried farming, selling real estate, and running for county engineer. He ended up a clerk in the family leather store in Galena, Illinois. The two men met briefly on a street in St. Louis in late 1857. Having not seen each other in sixteen years they compared notes on their lack of success in civilian life and agreed, as Sherman recalled, that "West Point and the Regular Army were not good schools for farmers, bankers, merchants, and mechanics."

The Army on the Eve of War

By December 1860, the Army's authorized strength totaled about 18,000 officers and men, but only 16,367 were on the rolls. Of these, 1,108 were commissioned officers, four were general officers (one major general who served as the commanding general and three brigadier generals), and the rest were either line officers assigned to the regiments or staff officers serving in the War Department. There were 361 staff officers assigned to the nine bureaus and departments, all of which were headed by colonels, although several held staff brevets of brigadier general. The bureau chiefs were men of long service, averaging sixty-four years of age, with six over seventy. The 743 line officers served in the regiments: 351 in the infantry, 210 in the artillery, and 182 in the mounted units. As with the bureau chiefs in the War Department, the nineteen regimental colonels were mostly old men set in their ways. They ranged in age from forty-two to eighty, the average being sixty-three.

The officers in each regimental headquarters consisted of a colonel, a lieutenant colonel, two majors, an adjutant, and a quartermaster. The adjutant and quartermaster were lieutenants detailed from the line companies except in the mounted regiments which were authorized additional lieutenants for the headquarters. The enlisted staff included a sergeant major, a quartermaster sergeant, and a chief musician. Infantry and artillery regiments were each authorized twenty musicians while mounted regiments had two chief buglers. Because Congress in 1850 had authorized units in the West to receive more men than those in the East, regimental strength could vary significantly. Theoretically a regiment in the West could have as many as nine hundred soldiers, but no regiment ever reached that size. With recruitment and desertion being perennial problems, a typical regiment averaged 300 to 400 enlisted men with 1 or 2 officers and 30 to 40 men in each company.

With ten infantry and five mounted regiments of ten companies each (the 8th Infantry had only nine) and four artillery regiments of twelve companies each, there were a total of 197 line, or combat, companies in the United States Army on the eve of the Civil War. Of these, only eighteen, all artillery, were stationed east of the Mississippi River.

As Southern states started to secede from the Union in the winter of 1860–1861 following Abraham Lincoln's victory in the presidential election, secessionists seized control of most Federal arsenals and forts in the South but allowed the officers and men to move north. One significant exception was the island bastion of Fort Sumter located in Charleston Harbor, South Carolina. President Lincoln's refusal to surrender the fort and the subsequent secessionist attack on it marked the formal outbreak of the Civil War.

The regular regiments responded well to the crisis. They assembled their far-flung detachments, marched east to join the fight, and, although miniscule in number compared with the more than seventeen hundred state volunteer regiments that served in the Union Army during the war, acquitted themselves well. But while the enlisted ranks remained solidly loyal, the Regular Army officer corps was more fractured. Capt. Charles Morton's description of the situation facing the officers in Regiment of Mounted Riflemen in late 1860 typified the dilemma of the officer corps: "One-third of our people had plunged into secession believing it right, another third declaring coercion wrong, but the other third taking the stand that saved the Union." Some of the officers "imbibed the epidemic political heresy of 'States' Rights,' and at no little sacrifice, cast their lot with the seceded States, breaking

Some Army installations, like Fort Moultrie, Charleston, South Carolina, were located in states that began seceding from the Union. (Accessible Archives)

close, tender and cherished ties of comradeship, and severing their connection with a service they revered and honored." Among those who "cast their lot" with the Confederacy were 168 West Point graduates. Of the other graduates in the Army, 556 remained loyal to the Union while 26 took no active part in the war. About 20 percent of the officer corps as a whole resigned from active service in 1861 to join the Confederate forces. As for those U.S. Military Academy graduates who had left the Army like Grant, Sherman, and Bragg, nearly 200 returned to don a uniform—92 wore Confederate gray and 102 put on the blue of the Union Army.

Regardless of which side they chose, Regular Army officers of the antebellum Army would have a disproportionate effect on the war. By the end of the war, 217 of the 583 men who had achieved general officer rank in the Federal Army were West Point graduates. One hundred forty-six of the Confederacy's 425 general officers were West Point graduates. But their impact went beyond that. According to Civil War historian T. Harry Williams, "Of the sixty biggest battles, West Point graduates commanded both armies in fifty-five, and in the remaining five a West Pointer commanded one of the opposing armies." Leavened on the frontier and in Mexico, the men of the antebellum officer corps would largely acquit themselves well in the greatest struggle the nation had known up to that point.

The Army of 1860, although woefully undermanned, was a much more effective force than the one that had marched into Mexico just fifteen years earlier. It had better weapons and a more professional officer corps. But like the Army of 1845, it would soon face a conflict for which it was unprepared. After the Mexican War, during a decade and a half of conducting small, independent operations on the frontier, junior officers had gained considerable experience in leading men in combat, but virtually none of the Army's senior leadership had ever fought a major battle. Men who had commanded companies of one hundred or fewer soldiers soon found themselves leading brigades, divisions, and corps. Even more so than in the Mexican War, the Army was about to undergo a stunning transformation as the nation rushed pell-mell toward mobilizing what would eventually become a million-man force of citizen soldiers who would fight to preserve the Union.

The Author

Clayton Newell is a retired Army officer who completed his career as chief of the historical services division at the U.S. Army Center of Military History. He now works as an independent scholar and military history consultant. Newell is the author of several books, including *Lee vs. McClellan: the First Campaign (2010)* and *Of Duty Well and Faithfully Done: A History of the Regular Army in the Civil War* (2011).

Further Readings

Ball, Durwood. *Army Regulars on the Western Frontier, 1848–1861*. Norman: University of Oklahoma Press, 2001.

Eisenhower, John S. D. *Agent of Destiny: The Life and Times of General Winfield Scott*. New York: The Free Press, 1997.

McCaffrey, James M. *Army of Manifest Destiny: The American Soldier in the Mexican War, 1846–1848*. New York: New York University Press, 1992.

Ness, George T. Jr., *The Regular Army on the Eve of the Civil War*. Baltimore, Md.: Toomey Press, 1990.

Utley, Robert M. *Frontiersmen in Blue: The United States Army and the Indian, 1848–1865*. New York: Macmillan, 1967. Reprinted Lincoln: University of Nebraska Press, 1981.

For more information, please read other titles in the series The U.S. Army Campaigns of the Mexican War and The U.S. Army Campaigns of the Civil War published by the U.S. Army Center of Military History (www.history.army.mil).